P9-BBO-916

Our Family Tree

AN EVOLUTION STORY

Lisa Westberg Peters

ILLUSTRATED BY Lauren Stringer

Harcourt, Inc.

San Diego New York London

Text copyright © 2003 by Lisa Westberg Peters
Illustrations copyright © 2003 by Lauren Stringer

All rights reserved. No part of this publication may be reproduced
or transmitted in any form or by any means, electronic or mechanical,
including photocopy, recording, or any information storage and
retrieval system, without permission in writing from the publisher.

For information about permission to reproduce selections from this book, write to
trade.permissions@hmhco.com or to Permissions, Houghton Mifflin Harcourt
Publishing Company, 3 Park Avenue, 19th Floor, New York, New York 10016.

www.hmhco.com

Library of Congress Cataloging-in-Publication Data
Peters, Lisa Westberg.
Our family tree: an evolution story/Lisa Westberg Peters;
illustrated by Lauren Stringer.
p. cm.
Summary: Relates the evolution of the family of mankind, from single cells
in the sea to human beings with "big brains that wonder who we are."
1. Evolution—Biology—Juvenile Literature. [1. Evolution—Fiction.]
I. Stringer, Lauren, ill. II. Title.
QH367.1.P48 2003
576.8—dc21 2002007806
ISBN 978-0-15-201772-9

SCP 17 16
4500603667

Printed in China

The illustrations in this book were painted in Lascaux acrylics
on Fabriano 140 lb. watercolor paper.
The display type was set in Minister Light.
The text type was set in Schneidler Medium.
Color separations by Bright Arts Ltd., Hong Kong
Printed and bound by RR Donnelley, China
Production supervision by Sandra Grebenar and Ginger Boyer
Designed by Lydia D'moch and Lauren Stringer

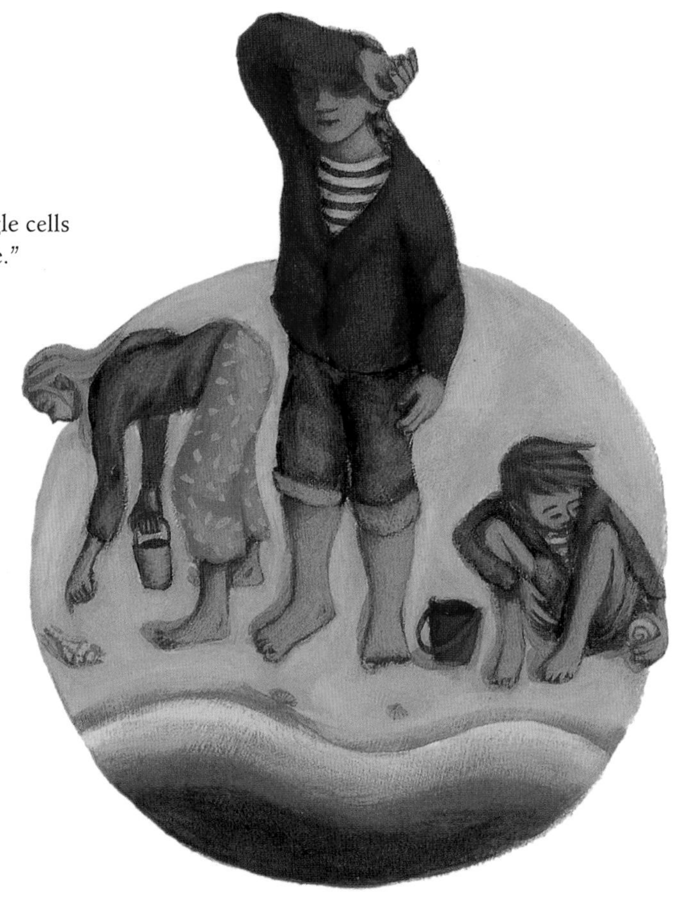

In memory of Steve Peaslee
—L. W. P.

For my family
—L. S.

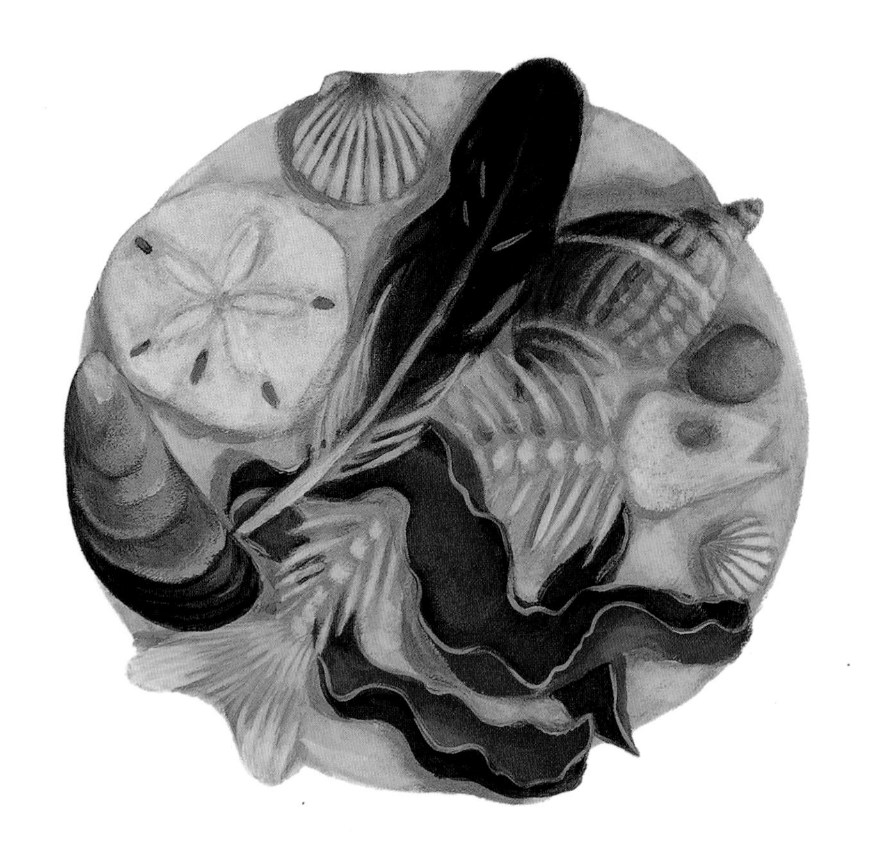

All of us are part of an old, old family.
The roots of our family tree
reach way back
to the beginning of life on earth.

We've changed a lot since then.

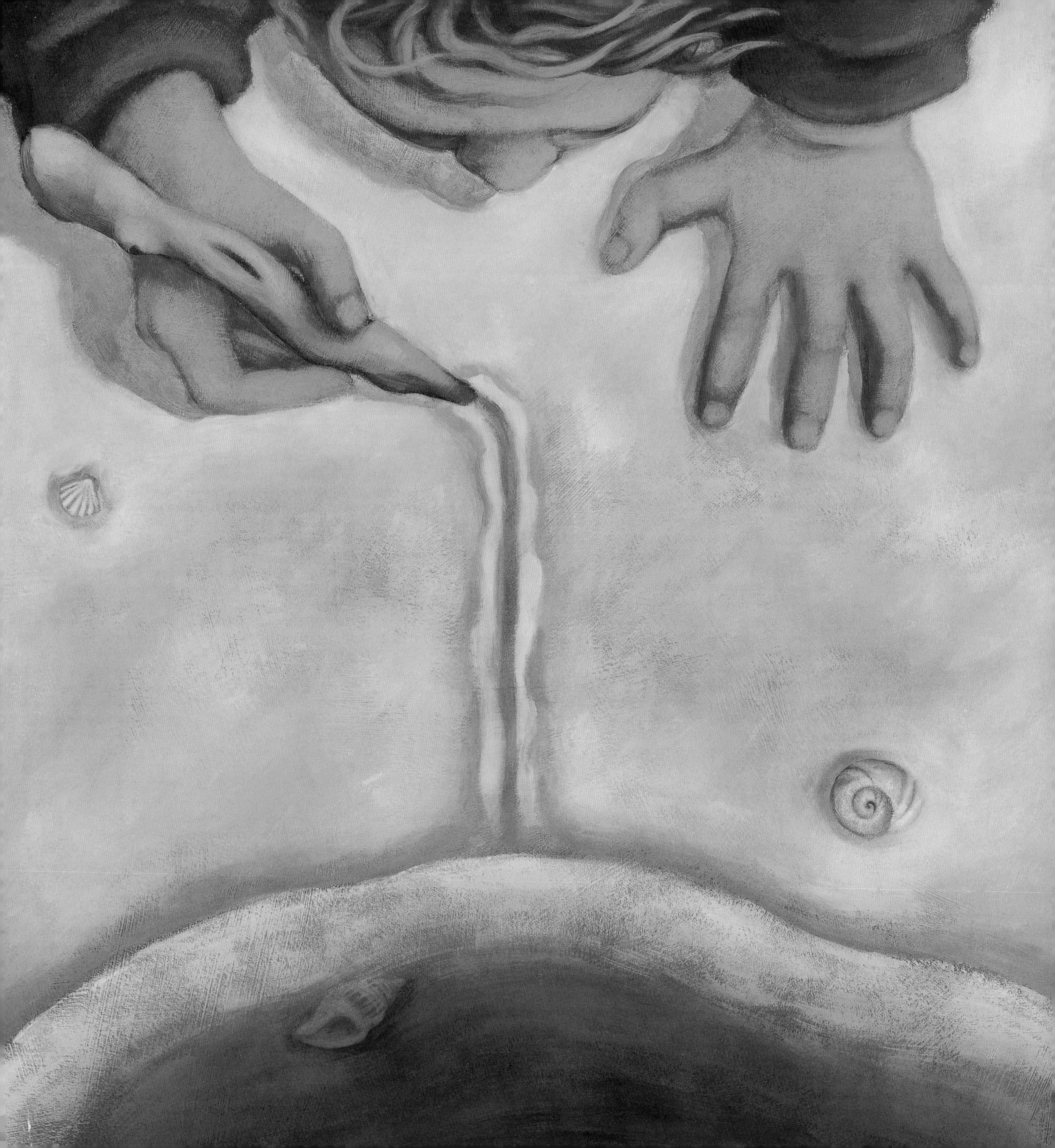

When we began,
we didn't look like people.
We didn't have two eyes to blink
or ten toes to wiggle.
We were just tiny round cells
in the deep, dark sea.

On the outside,
we were so small,
we were almost invisible.
But on the inside,
we had the same kind
of spiraling genetic code for life
we have today.

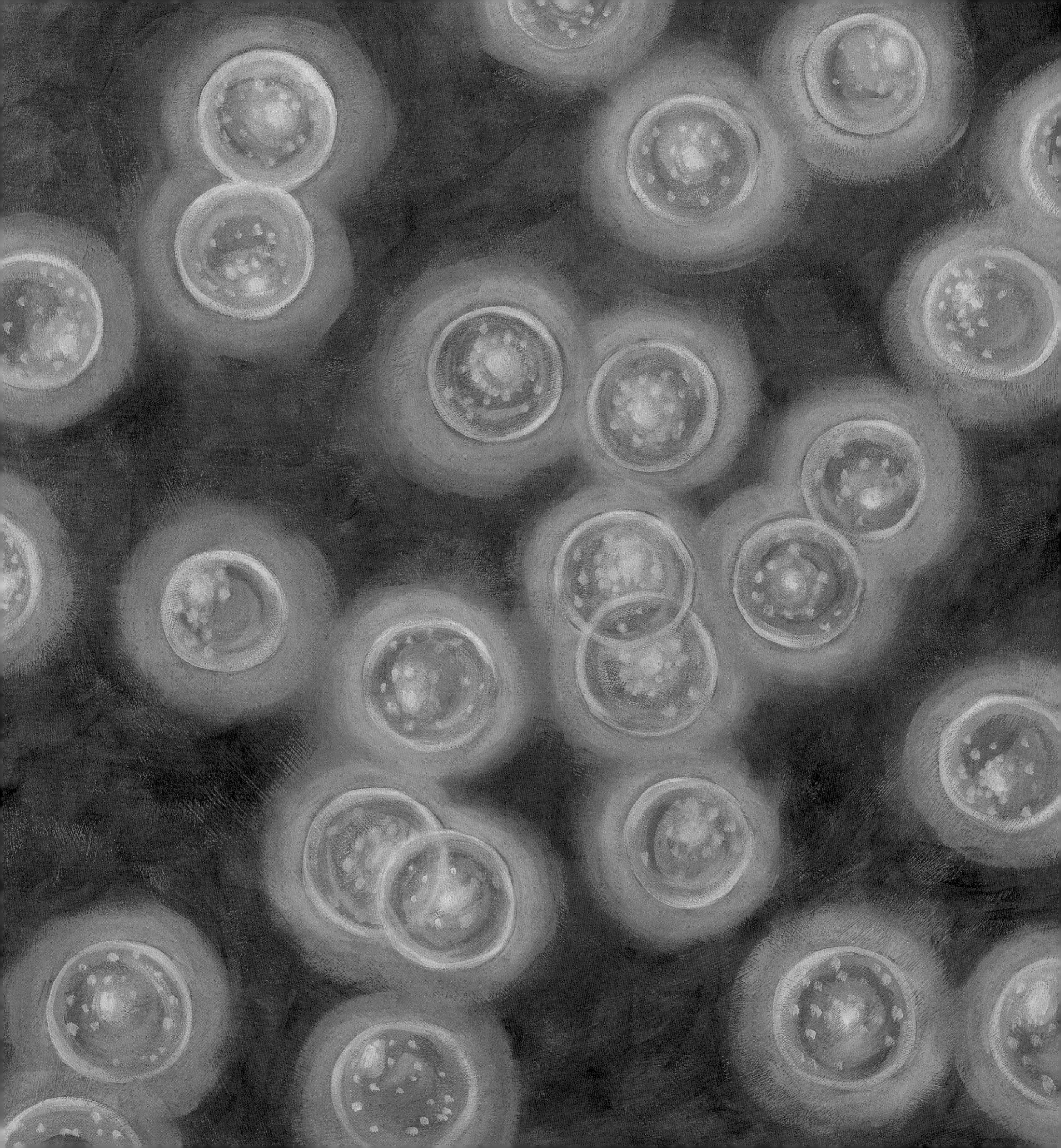

And that's the way our family stayed—
generation after generation,
year after year—
for millions of years:
tiny and round,
floating in the sea.

But then the earth changed.
Land rose from the oceans.
The air filled with oxygen.
Life changed, too.
Slowly...
slowly...
one step at a time,
some cells joined together,
and became plants.
Our cells joined together,
and we became animals.

On the outside,
we were squishy and soft,
like worms.
On the inside,
our cells had many shapes—
square like boxes,
pointy like stars,
round like ripe seeds—
the same way they do now.

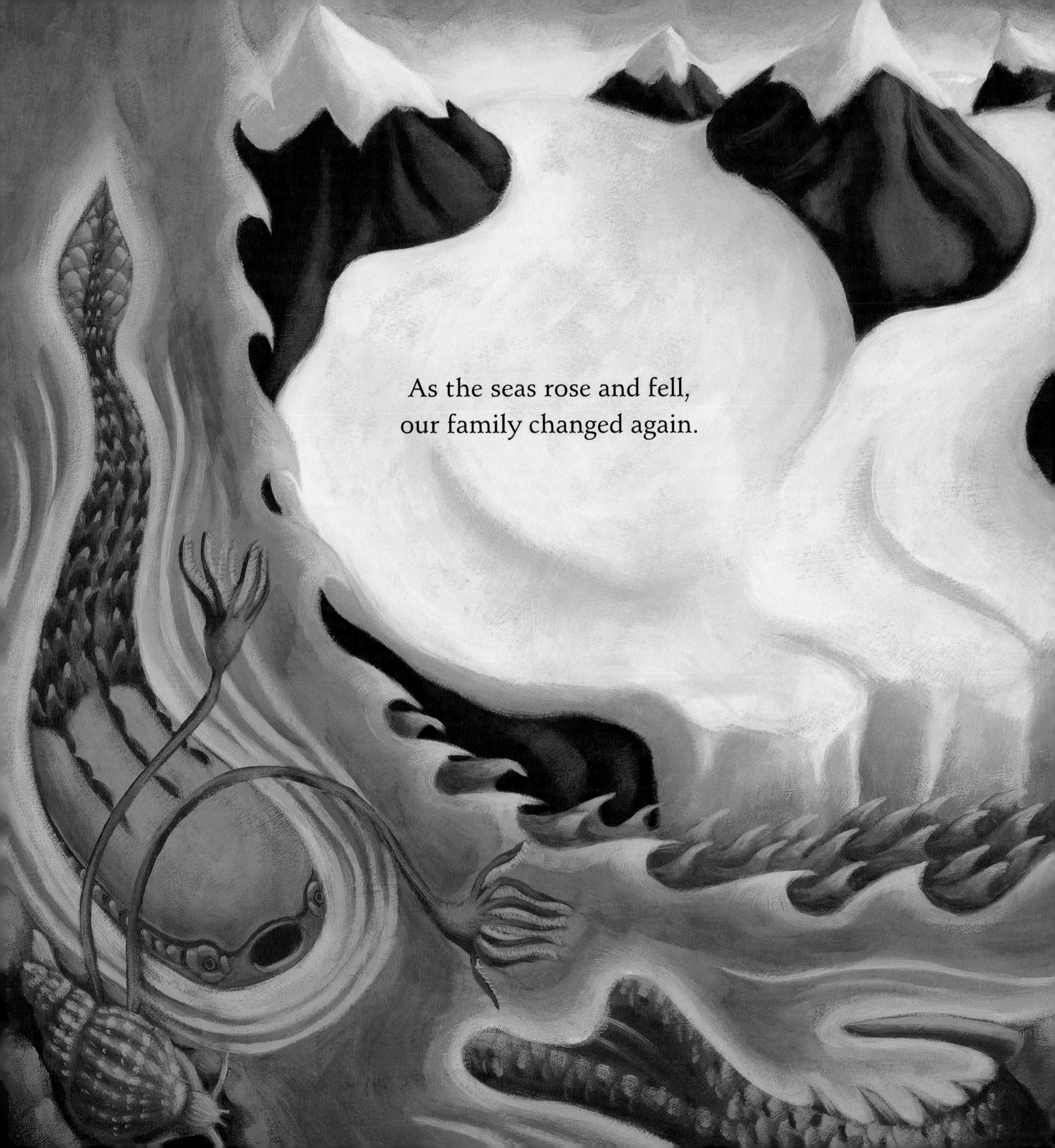

As the seas rose and fell,
our family changed again.

On the outside,
we had scales to protect us,
and fins—two on each side—
to swim against strong currents.
On the inside,
we had spiny backbones
that helped us move
as freely as we do
today.

When families of green plants and insects
began living on the land,
we followed them.

On the outside,
we still had scales,
but we looked like huge salamanders.
We crawled out of the sea on legs,
two on each side.
We didn't stray far,
and we returned to the water
to lay our eggs.
On the inside,
we had lungs to breathe oxygen,
like we do now.

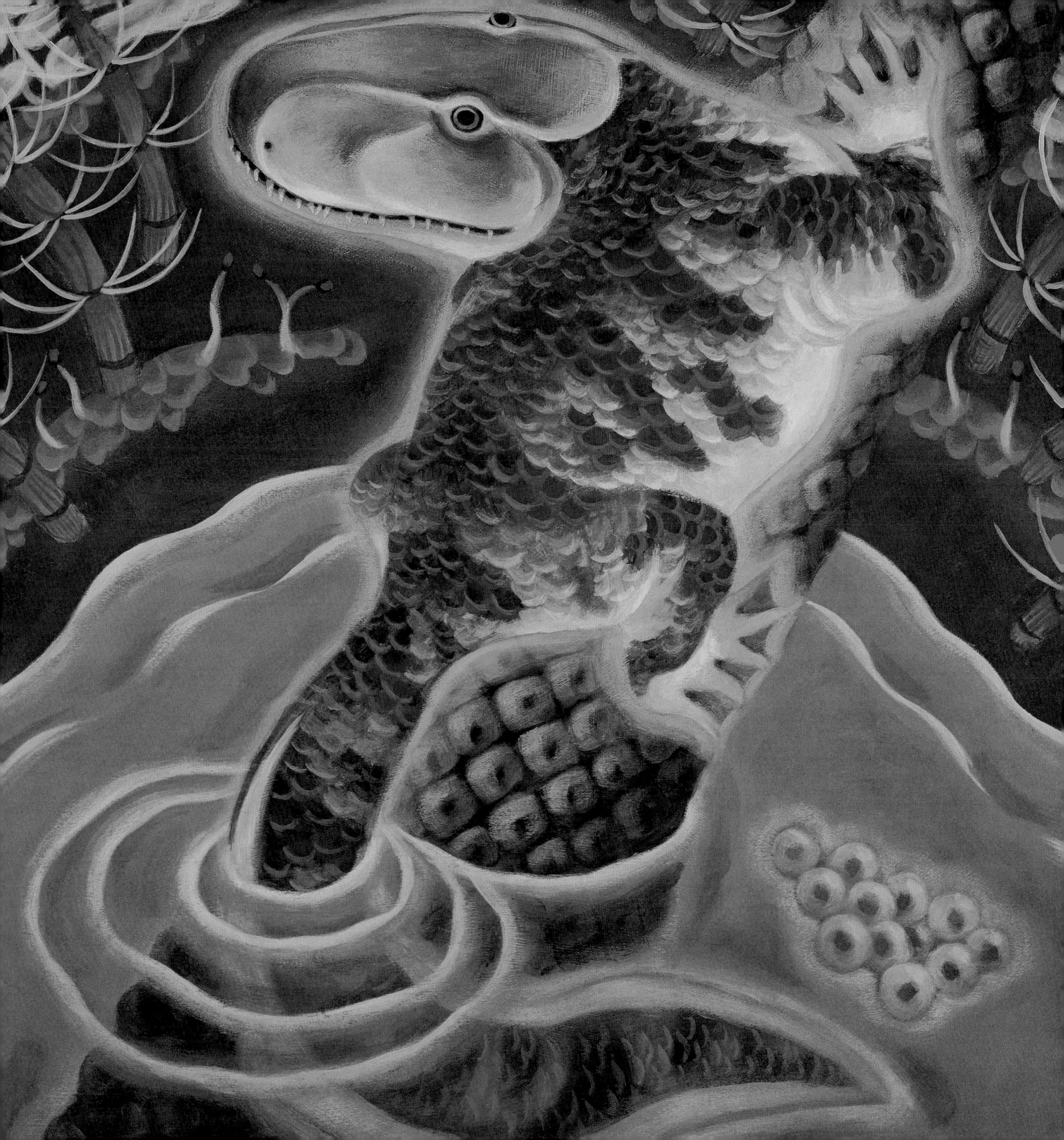

As all of the continents on earth
slowly joined into one,
we left the water completely.

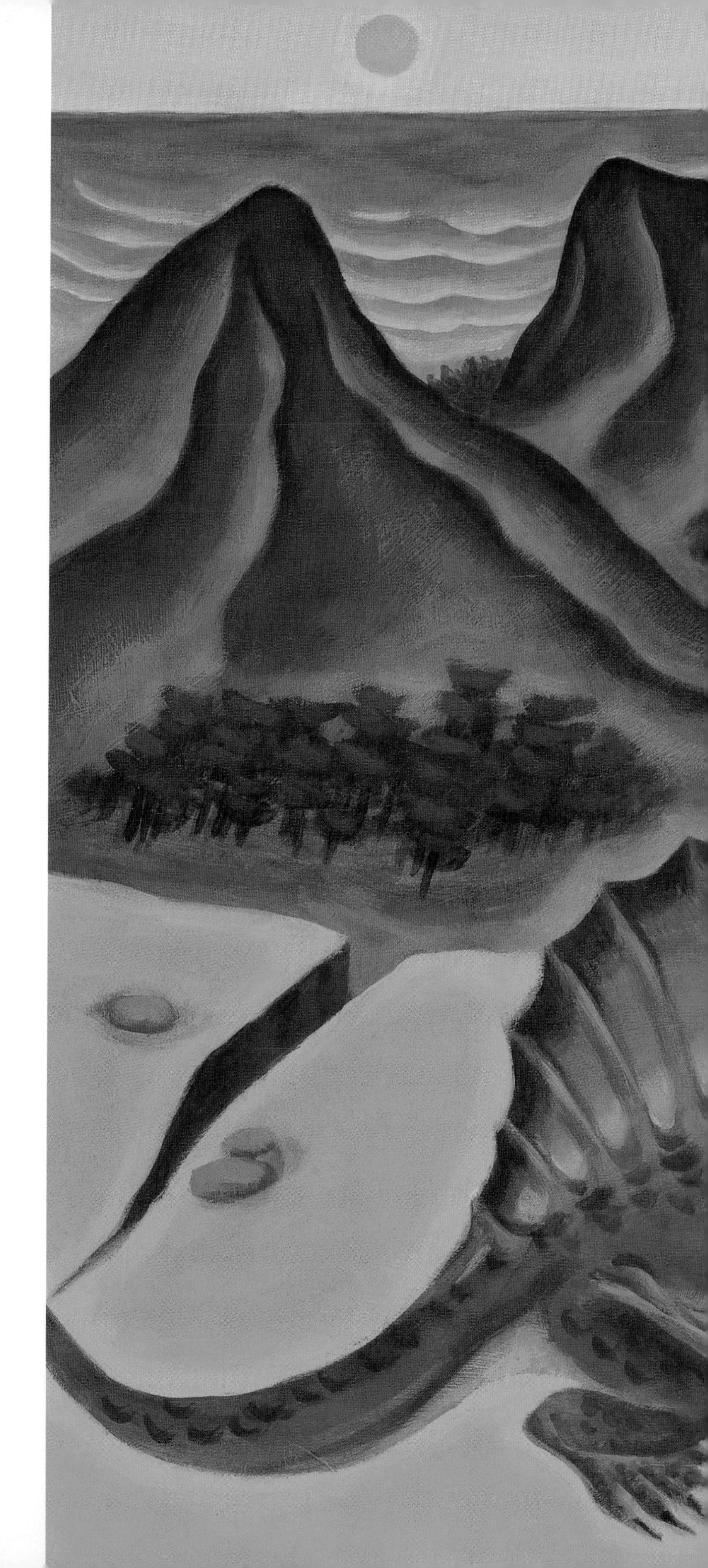

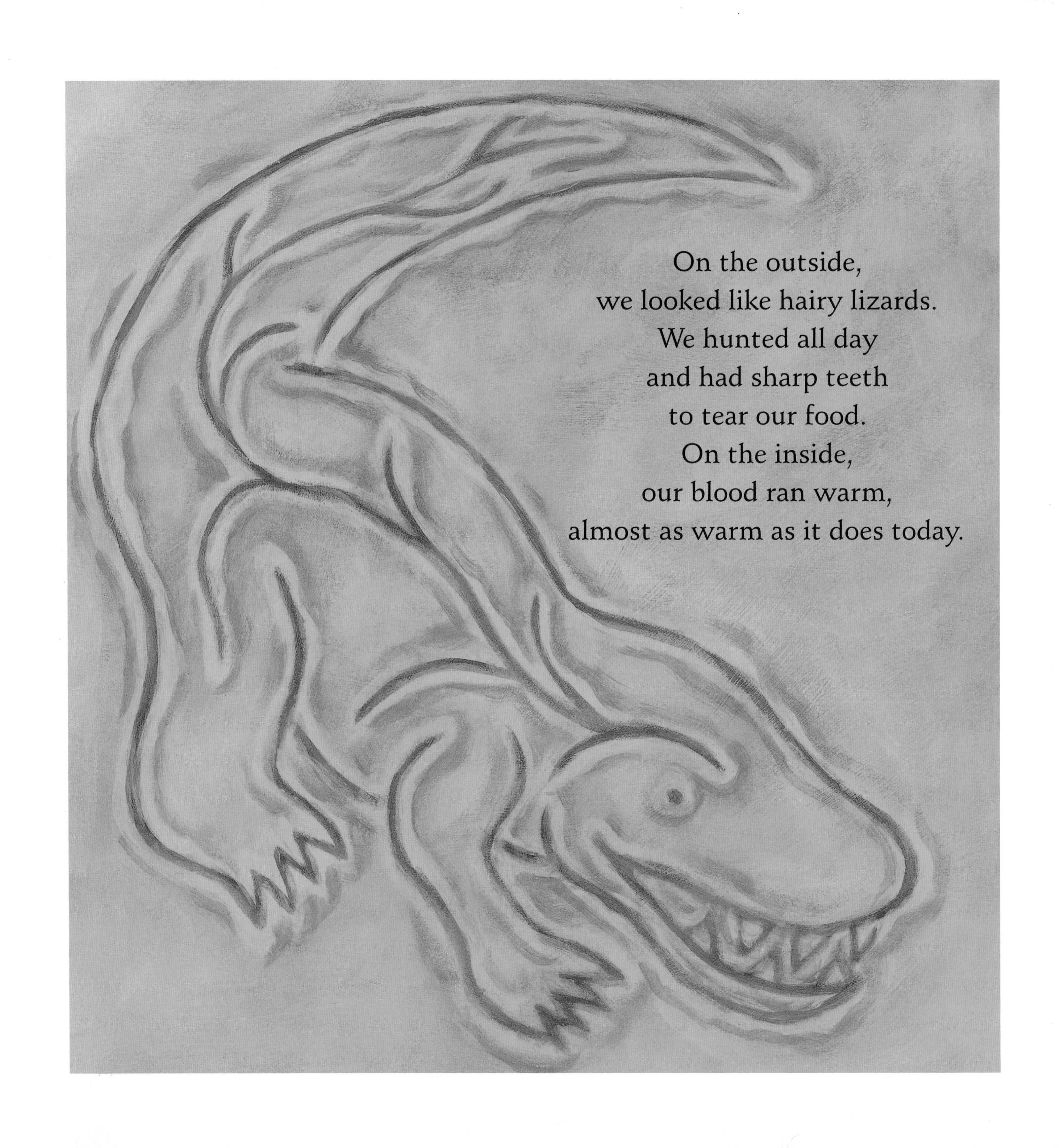

On the outside,
we looked like hairy lizards.
We hunted all day
and had sharp teeth
to tear our food.
On the inside,
our blood ran warm,
almost as warm as it does today.

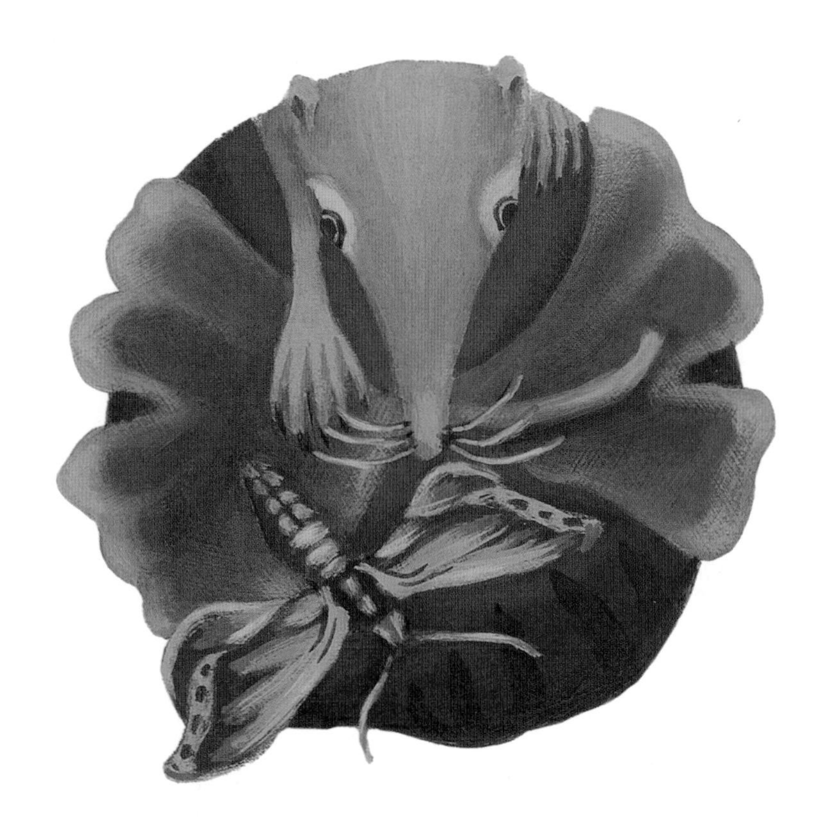

In one of the earth's dark times,
nearly all life went extinct,
but many families survived.
Ours was among them.

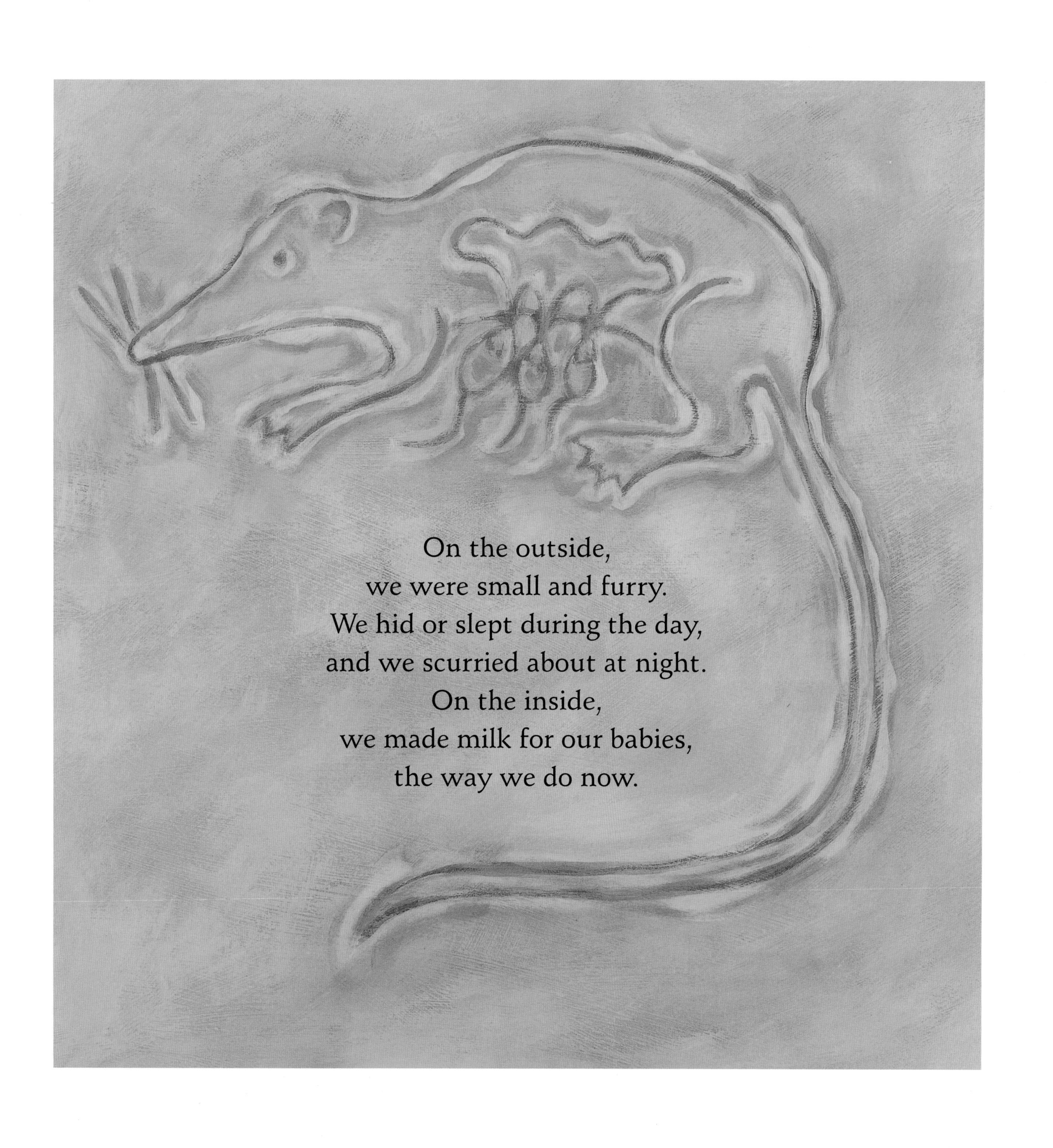

On the outside,
we were small and furry.
We hid or slept during the day,
and we scurried about at night.
On the inside,
we made milk for our babies,
the way we do now.

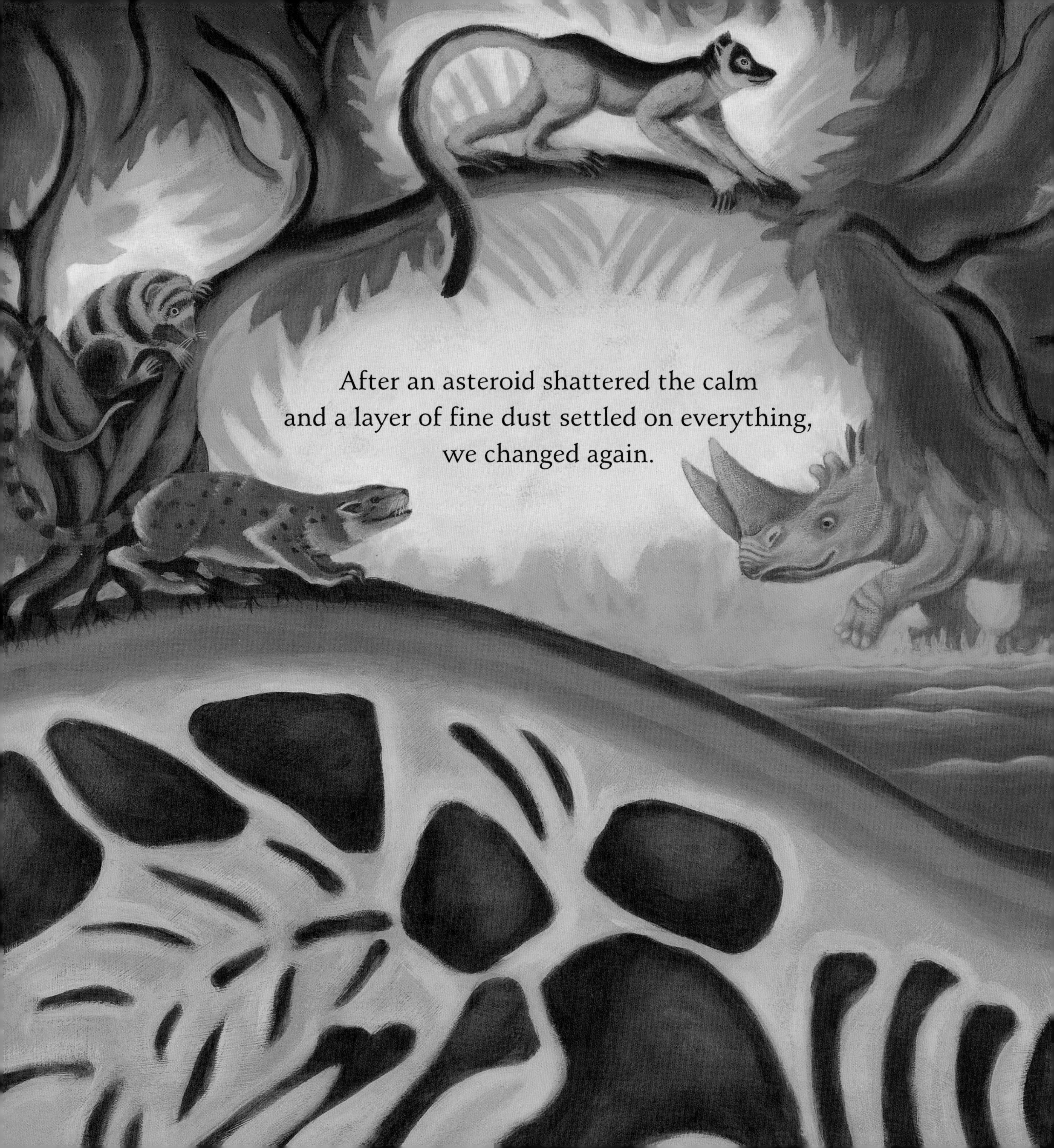

After an asteroid shattered the calm
and a layer of fine dust settled on everything,
we changed again.

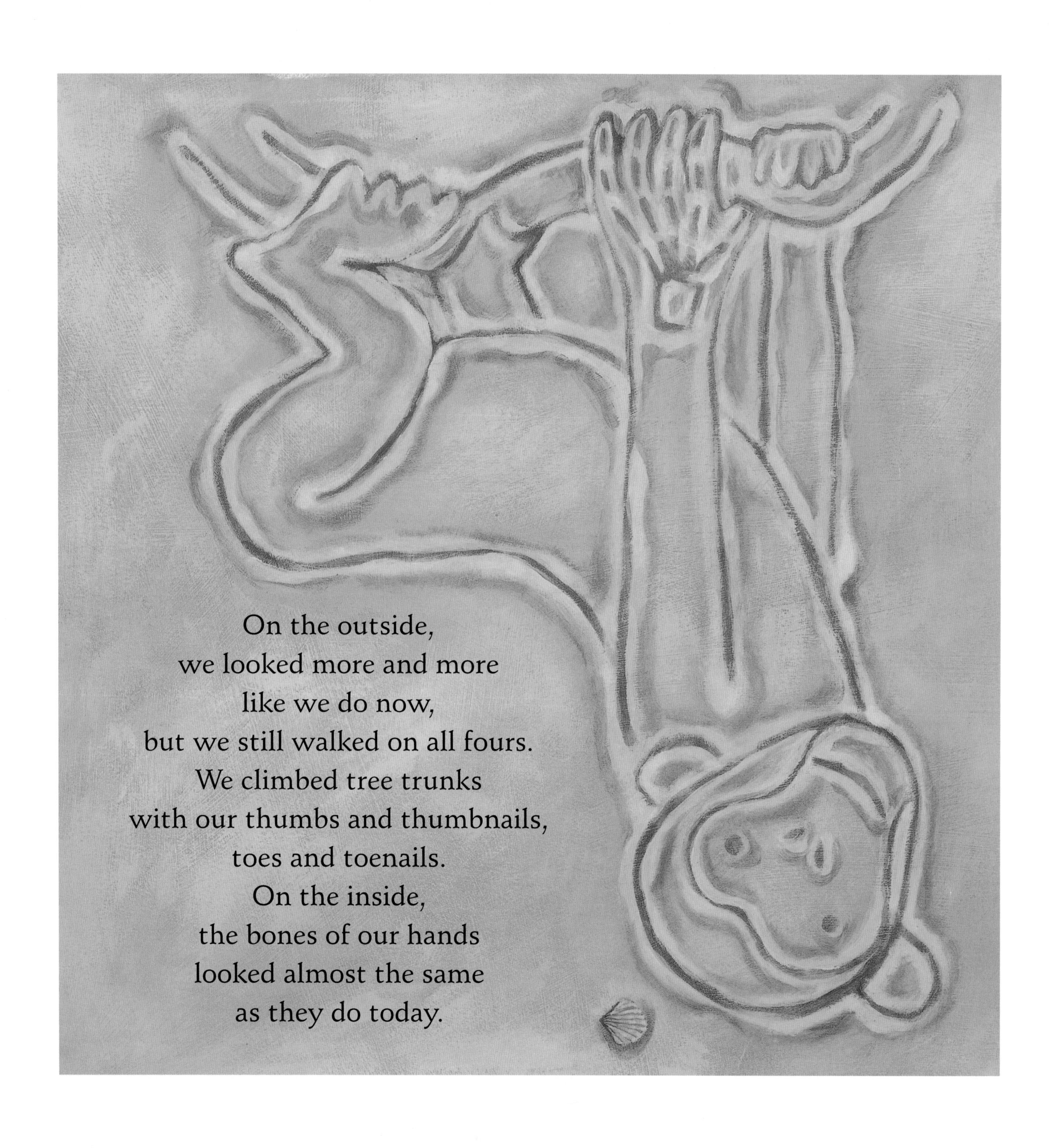

On the outside,
we looked more and more
like we do now,
but we still walked on all fours.
We climbed tree trunks
with our thumbs and thumbnails,
toes and toenails.
On the inside,
the bones of our hands
looked almost the same
as they do today.

After the earth cooled
and the forests shrank,
we left the trees
to live in the open grasslands.

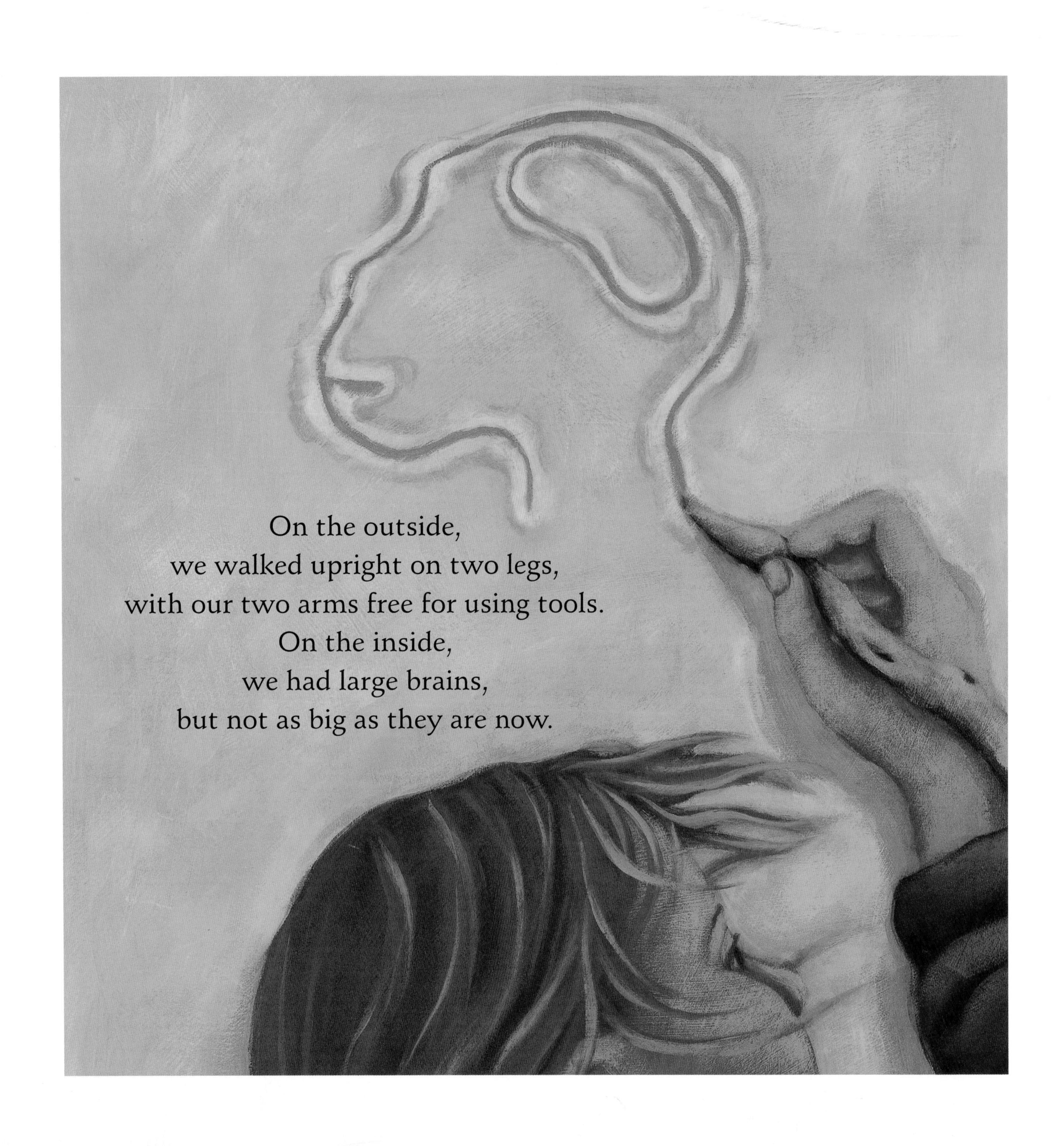

On the outside,
we walked upright on two legs,
with our two arms free for using tools.
On the inside,
we had large brains,
but not as big as they are now.

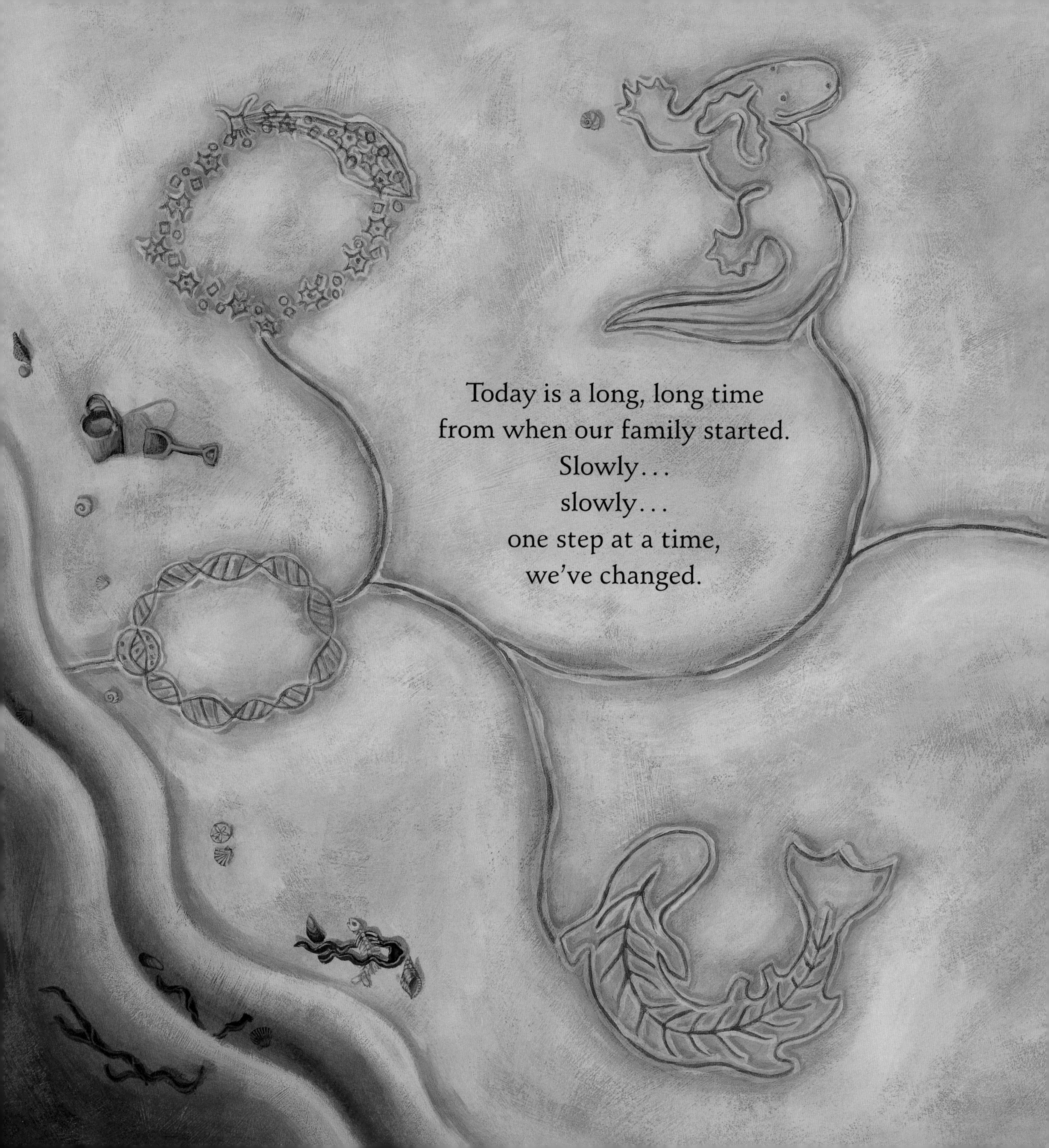

Today is a long, long time
from when our family started.
Slowly...
slowly...
one step at a time,
we've changed.

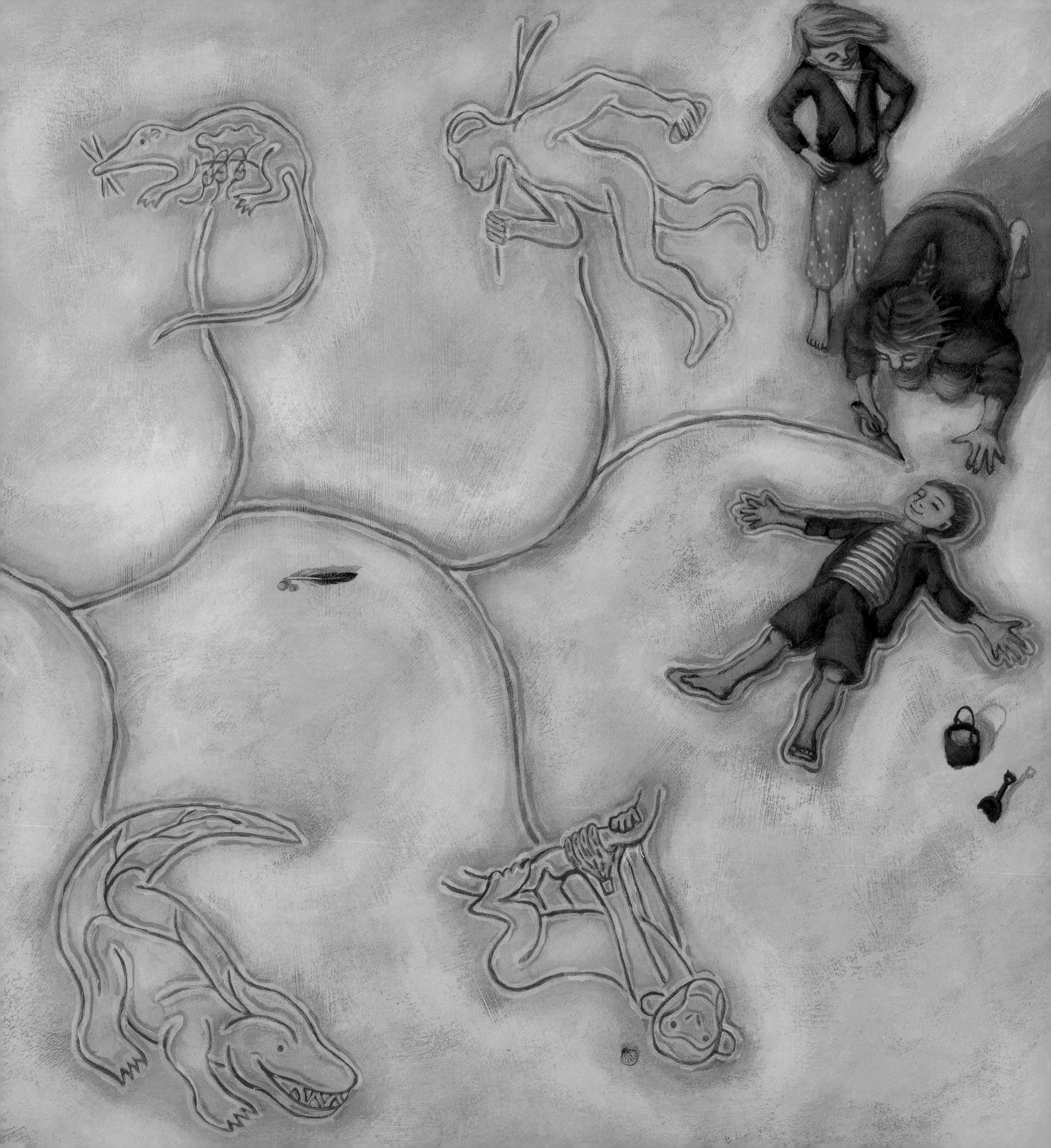

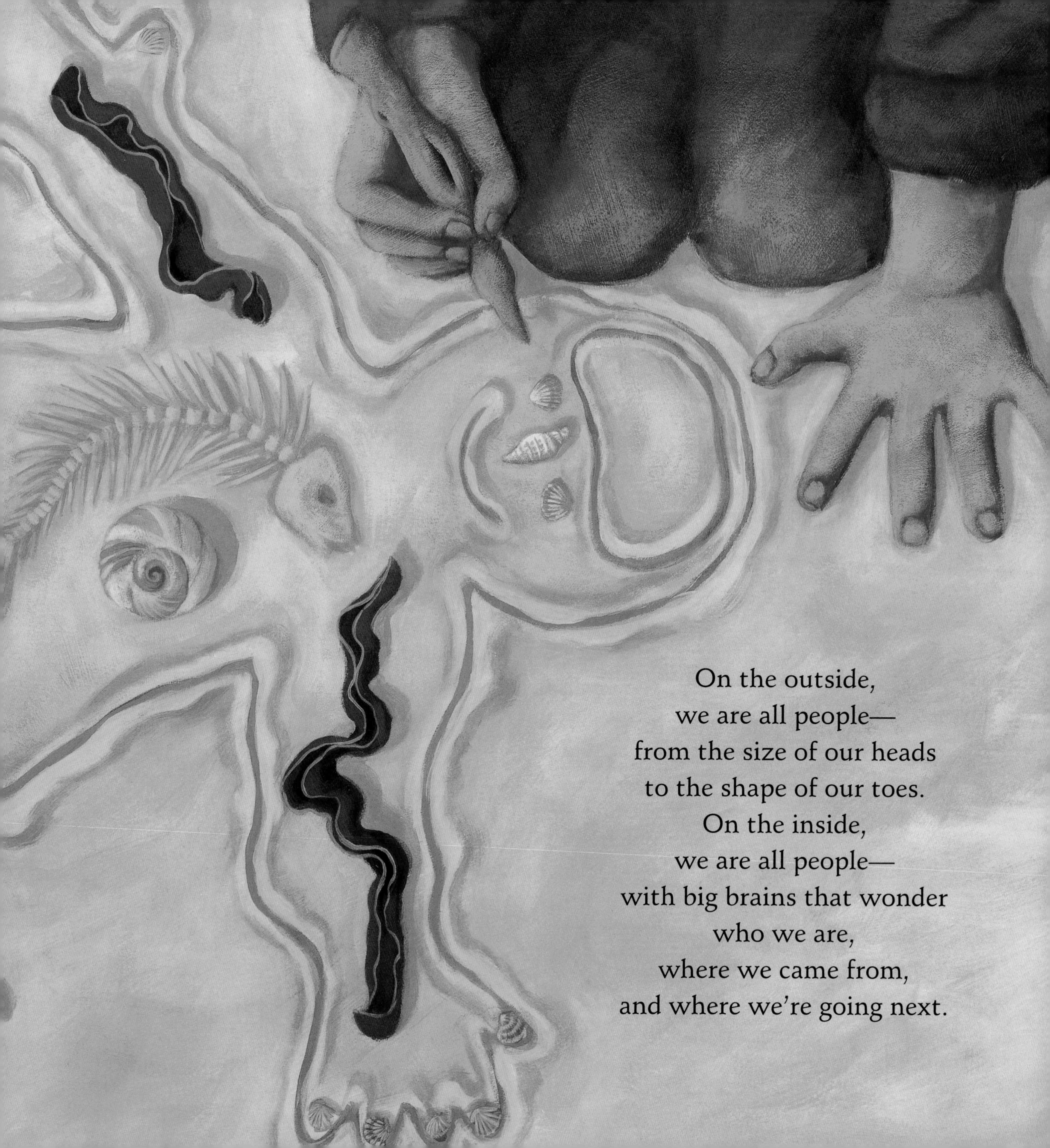

On the outside,
we are all people—
from the size of our heads
to the shape of our toes.
On the inside,
we are all people—
with big brains that wonder
who we are,
where we came from,
and where we're going next.

We began as tiny round cells,
and we've changed a lot since then.
But we carry with us reminders
of each step of our past.

That's how it is with families.
And ours goes back a long, long way.

The Roots of Our Family Tree

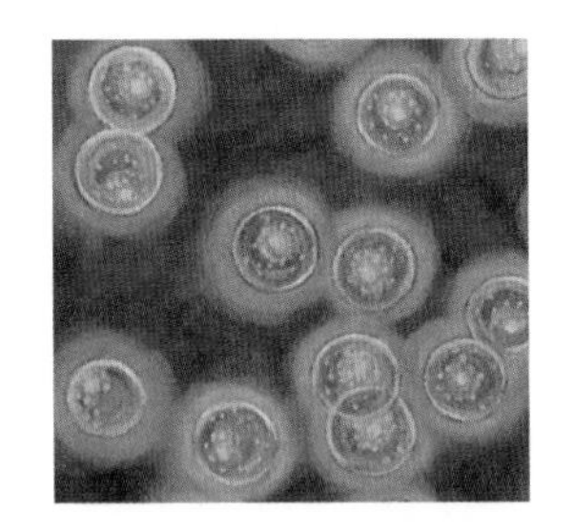

"We were just tiny round cells in the deep, dark sea."

We inherited something very basic from the first life on earth: a molecule with a long name—deoxyribonucleic acid, or DNA. DNA was the genetic code for these early living cells. When they reproduced, they passed on their genetic code to the next generation . . . and the next and the next. Today, all life still contains DNA, which means that all life is related and can be traced to a common ancestor. Single-celled life remains a very successful—and perhaps the dominant—form of life on earth.

"Our cells joined together, and we became animals."

We inherited specialized cells (like those for skin, muscles, and nerves) from the first animals. Specialized cells do different jobs, but they cooperate with each other. We also inherited the ability to use oxygen to extract energy from our food. Algae and plants don't consume oxygen; they produce it. (We split off from their family tree long ago.)

" . . . we had spiny backbones that helped us move as freely as we do today."

We inherited our backbones from the first fish. A backbone provides an attachment point for muscles and holds a cord inside for carrying messages to and from the brain. (Animals like lobsters and insects that have hard skeletons on the outside of their bodies are in a different family tree.)

" . . . we had lungs to breathe oxygen, like we do now."

We inherited our air-breathing lungs from fish, too. All fish use their gills to pull oxygen from the water, but some fish also developed lungs for gulping air. Some of these fish, the lobe-fins, began to explore the shorelines, "walking" with their strong fins. Over many, many generations, their four fins evolved into the four legs and the four feet of the earliest amphibians.

" . . . our blood ran warm, almost as warm as it does today."

We inherited our warm-bloodedness from a mammal-like reptile that is now extinct. These reptiles—the therapsids—were the first land animals to stay active (hunting and escaping predators) for most of the day. Cold-blooded animals, then and now, have to rest during the day to soak up heat from the sun. We also inherited hair, which insulates us from the cold, from the therapsids. (Birds, with their feathered insulation, evolved from dinosaurs and are in a different family tree.)

". . . we made milk for our babies, the way we do now."

Our mothers inherited their ability to produce milk for their babies from early mammals. Milk glands evolved from the sweat glands of these early mammals, which were as small as shrews and had to hide from the dinosaurs. These early mammals slept during the day and hunted for their food at night. They didn't develop larger bodies until the dominant dinosaurs went extinct sixty-five million years ago.

". . . the bones of our hands looked almost the same as they do today."

We inherited the five digits on our hands and feet from amphibians. We inherited our fingernails and toenails, which evolved from claws, from the early primates. These primates didn't need claws because they had opposable thumbs that could grip tree branches. The flat nails protected the tops of their fingers, and the sensitive pads on the opposite side helped the primates grasp insects and handle fruit.

". . . we had large brains, but not as big as they are now."

We inherited our larger-than-average (for mammals) brains from early primates, and most recently from early "hominids," our humanlike ancestors. The trend toward large brains may have persisted because these hominids had greater social demands. We also inherited our ability to walk upright from the early hominids. No one is entirely sure why these hominids started walking upright. One idea is that they found it advantageous to have their hands free for carrying things.

". . . we are all people—"

We inherited our distinctly human features from the first generations of *Homo sapiens,* which means "wise humans." *Homo sapiens* evolved 100,000 to 200,000 years ago in Africa. That's only about 7,000 generations ago—not very many considering how long life has existed on earth. Humans evolved so recently that any two of us are 99.9 percent identical, genetically. And yet, we are certainly different from each other. We inherit traits such as eye color and height—and maybe even some other things, like talents and diseases—from our most immediate ancestors, our parents.

Our Family Tree: A Timeline

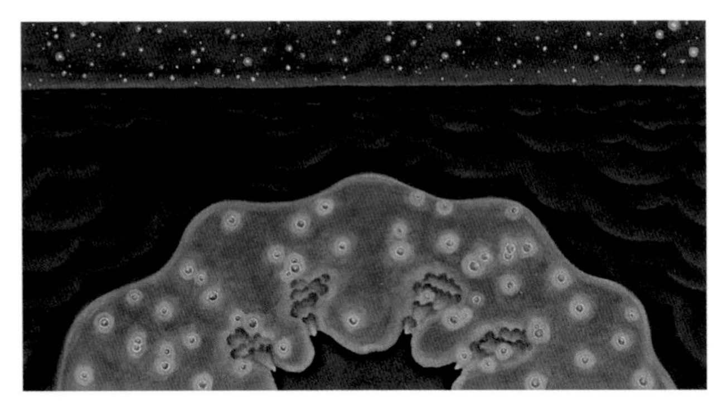

Life begins in the sea.

For much of earth's history, the only living organisms were microscopic bacteria and algae.

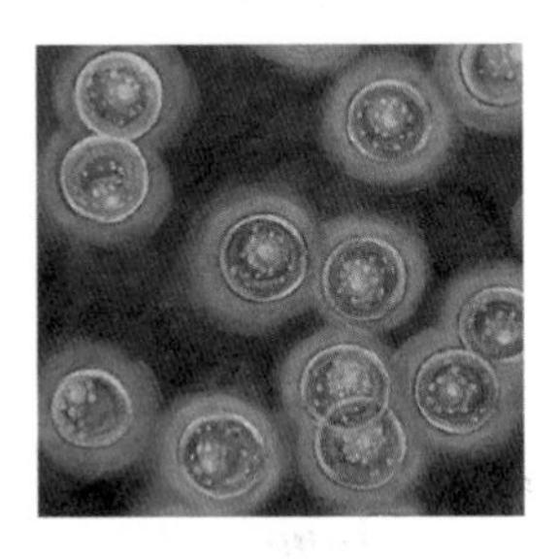

SINGLE CELLS

Sea levels rise and fall; glaciers come and go.

At least five major ice ages occurred throughout earth's history.

LOBE-FINNED FISH

3,800–3,600 | **520** | **400** | **370**

Millions of Years Ago

WORMLIKE VERTEBRATES

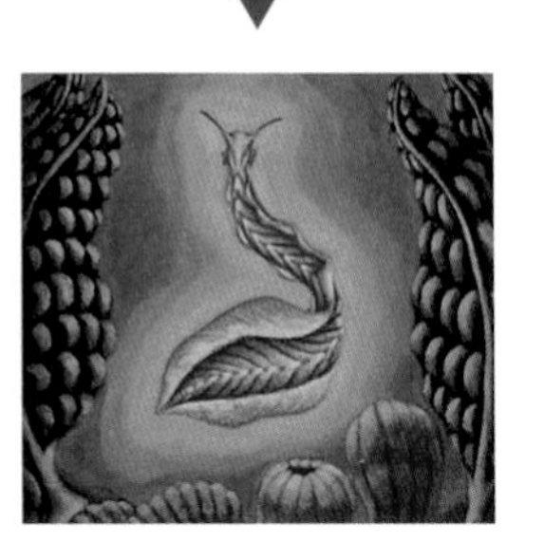

Oxygen accumulates in the sea and air.

Multicellular life took hold about 550 million years ago.

AMPHIBIANS

Life invades the land.

Plants and insects first appeared at least 450 million years ago.

Supercontinent Pangaea forms with deserts and mountains.

Pangaea existed from about 300–200 million years ago.

THERAPSIDS
(warm-blooded reptiles)

A mass extinction wipes out the dinosaurs, allowing mammals to thrive.

An asteroid impact 65 million years ago was the likely cause.

PRIMATES

MODERN HUMANS
(Homo sapiens)

260 **210** **55** **5–2** **.2–.1**

This timeline is not drawn to scale.

MAMMALS
(small, insect-eating)

The largest mass extinction in earth's history occurs 250 million years ago.

Ninety percent of all marine life died out.

EARLY HOMINIDS
(australopithecines)

Earth's climate is cooler and drier.

Grasslands became common about 20 million years ago.

Author's Note

My interest in the subject of evolution began with the writings of Stephen Jay Gould, who for years wrote a column for *Natural History* magazine. A sampling of his books for adults includes *Ever Since Darwin: Reflections in Natural History, Wonderful Life: The Burgess Shale and the Nature of History,* and *The Book of Life: An Illustrated History of the Evolution of Life on Earth.*

I also read Jared Diamond's excellent book *The Third Chimpanzee: The Evolution and Future of the Human Animal* (a *Los Angeles Times* Book Prize winner), and Jonathan Weiner's Pulitzer Prize winner, *The Beak of the Finch: A Story of Evolution in Our Time,* which is about the evolution of Darwin's finches on the Galapagos Islands.

As I began to delve into the subject, I read dense historical geology and evolution textbooks, but I also read a great children's book, *The Evolution Book* by Sara Stein. *The Beast in You! Activities and Questions to Explore Evolution* by Marc McCutcheon was also entertaining and informative.

The title essay in Barbara Kingsolver's book *High Tide in Tucson: Essays from Now or Never* was poetic inspiration.

The Internet wasn't widely used when I began working on my story, but I can now recommend several sites. The Smithsonian National Museum of Natural History site features a good discussion of human ancestors (www.mnh.si.edu/anthro/humanorigins). The University of California, Berkeley, Museum of Paleontology site presents a good overview of earth history and paleontology (www.ucmp.berkeley.edu). And the Institute of Human Origins at Arizona State University has a highly visual site that explores human evolution narrated by Donald Johanson, who discovered the "Lucy" hominid remains (www.becominghuman.org). The site also features a news section, a glossary, and good references.

I sought help from people as well as books. I'd like to thank them here: Dr. Barbara O'Connell, anthropologist, Hamline University, and Dr. Kent Kirkby, geologist, University of Minnesota, generously agreed to review the manuscript. Many other scientists—including Robert Sloan, Philip Regal, Robert Zink, Raymond Rogers, Andy Redline, David Fox, and Liz Goehring—answered questions along the way or read early drafts. Debra Frasier, Juanita Havill, Andrea Beebe, Nancy Olsen, and my Black Bear friends offered wisdom and advice. The Geological Society of Minnesota delighted me with years of lectures. The staff of the Saint Paul Public Library and the Ramsey County Library fetched books. Dave and the girls surrounded me with family. And my editor, Allyn Johnston, offered both passion for the subject and her unique editorial voice.

—L. W. P.

Illustrator's Note

My illustrations evolved from nearly two years of research done mostly at my local libraries. I purposely sought out heavily illustrated books to help me imagine what the first 3½ billion years of life on earth might have looked like. Two books that convey an amazing sense of the passage of time are *A Walk Through Time From Stardust to Us: The Evolution of Life on Earth* by Sidney Liebes, Elisabet Sahtouris, and Brian Swimme and *Rock of Ages, Sands of Time* by Barbara Page and Warren Allmon. Excellent detailed drawings can be found in both the *Atlas of the Prehistoric World* by Douglas Palmer and *The Simon & Schuster Encyclopedia of Dinosaurs & Prehistoric Creatures: A Visual Who's Who of Prehistoric Life* by Barry Cox, R. J. G. Savage, Brian Gardiner, Colin Harrison, and Douglas Palmer. And, finally, *Life: A Natural History of the First Four Billion Years of Life on Earth* by Richard Fortey was invaluable for its inspiring descriptions of our past.

Special thanks to Dr. Barbara O'Connell of Hamline University, for opening the doors of her anthropology lab and library to me, and also to Debra Frasier, who provided me with many beach and ocean pictures through a Minnesota winter. I am grateful to my editor, Allyn Johnston, for her patience and understanding that the evolution of a book can sometimes take longer than expected. I am forever thankful to my husband, Matthew Smith, whose support and critical eye were essential for me to complete the paintings. Finally, thank you to my children, Ruby and Cooper, my models for the beachcombing family, who knew just when to ask the most important questions.

—L. S.